AF784338

When looking up in the sky is the
inspiration for a book.

For all the stargazers big and small.

# What's that Twinkle in the Sky?

By Caroline J Knight

On a clear dark blue night,
have you ever looked up high,
and wondered to yourself,
"What's that twinkle in the sky?".

One twinkle, two twinkles,
three twinkles, four.

And there's a twinkle
I've never seen before!

There's one twinkle there,
two twinkles, three!
How far away are the twinkles
that I can see?

They're shiny and so bright,
the twinkles of each star.

The lights shining in the night sky
come from very far.

What is a star, why does it shine,
and why does it look so small?

Each star is light years away
and not so small at all!

Stars are massive balls of gas.
They're squeezed up very tight.

They fuse hydrogen and helium
producing heat and light.

While they spin and create energy,
the stars form into spheres.
They're held together by gravity
making light for millions of years.

spin spin spin spin

Stars come in many colours and types,
some are cool and some are hot.

Stars are yellow, blue or red, or white.
Some are bright and some are not.

There are spinning dense neutron stars,

and hot white dwarfs so bright.

Exhausted red giants swell right up,

and black holes steal the light.

Of all the stars we see in the sky,
there's one that's very close.

Our Sun! The star that makes our day!
The one we love the most.

# WHAT DOES THAT WORD MEAN ?

**Black Hole**
This kind of black hole is created after the collapse of a massive star. As it collapses in on itself, a very dense core is created. It has so much gravity that everything around it is pulled into it - even light! That's why it's called a black hole.

**Energy**
Is always changing from one form to another. Energy can be seen in the form of heat, light, something moving, electricity, a chemical reaction, and gravity.

**Gravity**
A force by which a body draws objects towards its centre.

**Helium**
A chemical element that contains two electrons, two proton and two neutrons. It is the second lightest and second most abundant chemical substance in the universe.

**Hydrogen**
A chemical element that contains one proton and one electron. It is the lightest and most common chemical substance in the universe.

**Neutron Star**
It's what is left when the core of a star has collapsed. Before it was a neutron star, it was a supernova which is an exploding star. Neutron stars are very dense and very small and spin very fast. They can also emit light as pulsars.

**Red Giant**
Is a dying star. This kind of star expands to a very large size after firstly running out of energy and collapsing in on itself to a point. It then grows to many times it's original size. It is able to start nuclear fusion again and it produces a lot of energy.

**Sphere**
A perfectly round object shaped like a ball.

**Star**
Massive ball of gas fused together and held together by gravity. It's made up of mostly hydrogen and helium and undergoes nuclear fusion producing heat and light.

**White Dwarf**
Is a star that has run out of its energy. It is very dense and very small. White Dwarfs slowly cool down over a very long period of time - billions of years.

Frankie & Harry's
BOOKSHELF

FRANKIE & HARRY'S BOOKSHELF
AUSTRALIA



Layout and cover font by Tim Cederwall
Editor : Lindsay Sant
Special thanks to Philip Knight

This edition published 2017



ISBN 978-0-6481459-2-9

www.frankieandharrysbookshelf.com

www.ingramcontent.com/pod-product-compliance
Lightning Source LLC
Chambersburg PA
CBHW040907311225
37518CB00035BB/989
* 9 7 8 0 6 4 8 1 4 5 9 2 9 *